TARSIERS AS POTENTIAL PETS

A COMPLETE GUIDE TO THEIR HABITAT, DIET, PRO'S AND CON'S, MANAGEMENT, BREEDING AND MANY MORE

DR HUNTER DAVIS

Copyright© 2024 **DR HUNTER DAVIS**

All rights reserved. No part or part of this book or publication may be reproduced, stored, or transferred in any form by electronic, mechanical, recording, or other retrieval system without written permission from the publisher

Table of Contents

INTRODUCTION 5

CHAPTER 1 7
    THE ALLURE OF PET TARSIERS 7

CHAPTER 2 15
    NEEDS FOR TARSIER HABITAT AND ENVIRONMENT 15

CHAPTER 3 24
    LEGALITIES AND ETHICAL ASPECTS 24

CHAPTER 4 32
    PET TARSIERS' NUTRITION AND DIET 32

CHAPTER 5 40
    TARSIER SOCIALIZATION AND BEHAVIOR 40

CHAPTER 6 50
    VETERINARY AND HEALTH CARE CONSIDERATIONS 50

CHAPTER 7 62
    ACTIVITIES FOR TRAINING AND ENRICHMENT 62

CHAPTER 8 75

Conscientious Ownership of Tarsiers 75

CHAPTER 9 89

In conclusion, would you be a good home for a tarsier? 89

CHAPTER 10 98

Commonly Asked Questions (FAQs) regarding Ownership of Tarsiers 98

Introduction

Many animal enthusiasts are enthralled with tarsiers because of their captivating large eyes and distinctive physical characteristics. These tiny, nocturnal primates are indigenous to a number of islands in Southeast Asia, where they live in thick jungles and forests. Although tarsiers are mainly recognized for their wild existence, there has been a growing trend of tarsier pet ownership. Nonetheless, the idea of taking care of these intriguing animals brings up significant questions about their welfare, status as legal animals, and suitability as pets. The appeal of owning tarsiers as pets is examined in this introduction, along with the moral and practical ramifications of doing so.

It's critical to comprehend tarsiers' natural behaviors, habitat needs, and dietary requirements as we delve deeper into the topic of tarsiers as possible pets. These small primates require special care because they have evolved to fill particular ecological niches, which can be difficult at times. Furthermore, the choice to maintain tarsiers in captivity is made more difficult by moral dilemmas pertaining to the pet trade and conservation initiatives. With thorough understanding of the intricacies of tarsier ownership, this investigation seeks to enable people to make knowledgeable decisions about bringing these fascinating animals into their homes.

Chapter 1

The Allure of Pet Tarsiers

With their remarkably large eyes and distinctive physical characteristics, tarsiers have long piqued the interest of animal lovers everywhere. These tiny, nocturnal primates, which are native to Southeast Asia's dense jungles and forests, have a unique quality that makes them stand out from other animal species. Although tarsiers have historically been thought of as wild animals, there is increasing interest in them as possible pets. The fascination of these fascinating animals and the difficulties of providing for and bonding with them are intriguingly called into question by this trend. We delve into the many facets of tarsiers' appeal as pets in this thorough examination, looking at their endearing qualities, their suitability for domestic life, and the moral dilemmas associated with owning one.

1: Tarsiers' Mysterious Allure

With their unusually shaped heads, long fingers, and disproportionately large eyes, tarsiers are known for their captivating appearance. These characteristics add to their ethereal, almost mystical appeal, along with their small stature and deft movements. Particularly designed for nocturnal hunting, tarsiers' large eyes allow them to detect prey with amazing accuracy even in low light. Their ability to hunt has been improved by this special adaptation, which also makes them more visually appealing and evocative of mystery and intrigue.

In addition, tarsiers exhibit a variety of fascinating behaviors that make them lovable to observers. One can only be in awe and admiration of their acrobatic agility and quick movements as they leap from branch to branch. Their vocalizations, which range from low chirps to high-pitched calls, also enhance their allure and mysterious aura. Seeing tarsiers in their natural environment and engaging in these activities provides an insight into their intriguing world and piques one's curiosity about these fascinating animals.

2: Tarsiers' Distinct Personality

In addition to their physical characteristics, tarsiers have a variety of fascinating personality traits that make them even more appealing as possible pets. Even though they are small in stature, tarsiers are intelligent and curious animals that frequently exhibit sophisticated social skills and problem-solving techniques in their family groups. Even though they spend most of their time alone in the wild, tarsiers can develop close relationships with humans who care for them, showing loving behaviors and an unexpected degree of adaptability to domestic settings.

In addition, tarsiers are renowned for their playful nature, stimulating their bodies and minds with a variety of enrichment and exploration activities. Tarsiers exhibit an adorable and entertaining zest for life, from exploring new objects to acting out hunting scenarios. They are fascinating companions for people looking for a dynamic and engaging pet experience because of their curious disposition and exploratory nature.

3: Tarsiers' exotic appeal as pets

Because they are unusual and unusual pets, tarsiers have an exotic charm all their own. Tarsiers, as opposed to more popular domestic animals like cats and dogs, have a unique and exotic presence that draws people looking for a unique pet ownership experience. Their rarity and elusiveness make them more appealing to enthusiasts who are lured to the idea of taking care of a species that the general public does not easily access or recognize.

In addition, having daily interactions with tarsiers fosters a sense of connectedness to the natural world, which in turn deepens appreciation for biodiversity and the significance of conservation efforts. Pet owners who welcome tarsiers into their homes serve as ambassadors for the species, bringing attention to the difficulties the animals face in the wild and the value of protecting their natural environments.

4: The Difficulties of Owning a Tarsier

Although keeping tarsiers as pets has obvious benefits, it's important to recognize the difficulties and obligations involved in taking care of these unusual animals. For the sake of their health and wellbeing, tarsiers have certain dietary and environmental needs that must be satisfied. Due to their nocturnal habits and sensitivity to light and noise, their living space must be modified. To reduce stress and discomfort, dim lighting and quiet surroundings must be provided.

In addition, tarsiers are extremely specialized feeders that, in the wild, mostly eat insects and small vertebrates. In order to ensure nutritional sufficiency, it can be difficult to replicate their natural diet in captivity and may need careful planning and supplementation. Furthermore, because tarsiers are gregarious creatures who enjoy company and stimulation, they need lots of opportunities for mental and enrichment activities to avoid boredom and behavioral problems.

5: Ethical Aspects of Owning Tarsiers

Important ethical questions concerning tarsiers' welfare and status as endangered species are brought up by the choice to keep them as pets. Some people contend that the exotic pet trade endangers wild populations and undercuts efforts to protect tarsiers in their natural habitats, while others see pet ownership as a way to promote conservation awareness and the preservation of species.

In addition, tarsier poaching and trade for the pet trade may negatively impact wild populations, resulting in population decreases and habitat degradation. In addition to advocating for the preservation of tarsiers in their native habitats, conservation organizations and animal welfare groups also support ecotourism and other initiatives that restore habitat.

Tarsiers are a popular choice for pets because of their captivating looks, gregarious nature, and exotic appeal. That being said, bringing a tarsier into one's house entails serious obligations and moral decisions. Pet owners can help ensure that these amazing animals are around for future generations to admire and appreciate by being aware of the challenges involved in caring for tarsiers and speaking out in favor of their welfare and conservation.

Chapter 2

Needs for Tarsier Habitat and Environment

Native to Southeast Asia's deep forests and jungles, tarsiers are amazing animals with distinctive features and an alluring appearance. Anyone thinking about keeping tarsiers as pets needs to be aware of their natural habitat and environmental needs. In this thorough investigation, we examine the components required to establish a suitable living environment for these fascinating primates in captivity, delving into the nuances of tarsier habitat and environmental requirements.

1. Tarsiers' Natural Habitat

Primarily found in Southeast Asia, tarsiers can be found in tropical forests and secondary growth habitats in countries like Sumatra, Borneo, Sulawesi, and the Philippines. These areas are distinguished by thick undergrowth, tall trees, and a wide variety of plant and animal species. Tarsiers live in the upper reaches of the forest canopy in their natural habitat. They use their nimble limbs and remarkable leaping skills to get through the dense foliage.

Depending on the species and region, tarsiers have different preferences for particular habitats. Certain species of tarsiers live in primary and secondary forests, like the Philippine tarsier (Carlito syrichta), while other species—like the spectral tarsier (Tarsius spectrum)—are found in disturbed habitats and agricultural areas. Notwithstanding these differences, tarsiers have similar needs for their habitat, such as access to enough food supplies, cover from the elements, and room for their territorial and foraging activities.

2: Tarsier Captive Environmental Conditions

For the health and welfare of tarsiers kept in captivity, it is imperative to recreate their natural environment. To make sure that the captive environment satisfies the tarsier's physiological and behavioral needs, a number of important environmental factors need to be carefully taken into account.

Temperature and Humidity: Because of their adaptation to tropical climates, tarsiers thrive in warm, humid environments. To keep pet tarsiers stress-free and comfortable, the captive enclosure's humidity and temperature must be kept at the proper levels. Ideal temperature ranges are from 24 to 28 degrees Celsius (75 to 82 degrees Fahrenheit), with humidity levels kept between 50% and 80%.

Lighting: Because they are nocturnal creatures, tarsiers are most active at night and have adapted to live in low light. To replicate the natural light cycle that tarsiers experience in the wild, the captive enclosure must have the proper lighting. You can use dim red lights or full spectrum lighting to create a nighttime habitat that is ideal for tarsier behavior and activity.

Enclosure Design: To facilitate climbing, jumping, and exploration, the enclosure design for tarsiers should imitate the vertical structure of their natural habitat. Enclosures should have several levels, branches, and platforms for climbing and perching in order to accommodate tarsier behaviors. For captive tarsiers, an environment that is both stimulating and enriching can be created by incorporating natural materials like branches, vines, and foliage.

3: Pet Tarsiers' Dietary Requirements

For pet tarsiers to be healthy and happy, proper nutrition must be provided in addition to considerations regarding their surroundings. The majority of tarsiers' diet in the wild consists of insects, spiders, and small vertebrates. It is essential to mimic their natural diet in captivity in order to guarantee adequate nutrition and avoid health problems associated with diet.

Feeding Schedule: Live insects like crickets, mealworms, and roaches should be the main source of nutrition for pet tarsiers. Commercial primate diets and small amounts of fruits and vegetables should also be provided. In order to accommodate their innate feeding habits, feeding should take place at night or in the evening.

Supplementation: To maintain nutritional balance, the tarsier's diet may need to be supplemented with extra vitamins and minerals. In particular, calcium supplements are critical for keeping strong bones and averting metabolic bone disease.

Water: To replicate natural dew collection, fresh water should always be available, either in a shallow dish or via a drip system. Although tarsiers can also get moisture from their diet, access to potable water is crucial for both hydration and general health.

4: Socialization and Behavioral Enrichment

Tarsiers are gregarious, highly intelligent animals that enjoy mental and social stimulation. To keep captive tarsiers from becoming bored and to encourage their natural behaviors, it is crucial to provide them with opportunities for behavioral enrichment and socialization.

Enrichment Activities: Foraging, exploring, and problem-solving should all be part of the enrichment activities. To pique the tarsier's curiosity and promote instinctive behaviors, include toys and novel objects, puzzle feeders, and hidden treats inside the enclosure.

Socialization: Although tarsiers spend most of their time alone in the wild, having conspecifics or compatible companions in captivity may be beneficial to them. To avoid aggression and guarantee compatibility amongst individuals, cautious supervision and gradual introductions are required when bringing multiple tarsiers into the same enclosure.

5: Veterinary Treatment and Health Surveillance

To keep pet tarsiers healthy and happy, regular veterinary care is necessary. Regular immunizations, fecal parasite testing, and annual wellness exams are advised to avoid and identify health problems early.

Signs of Illness: Owners of pets should be aware of the symptoms of tarsiers' illnesses, which include alterations in their appearance, behavior, and appetite. Reporting any strange symptoms to a licensed veterinarian with experience caring for exotic animals should happen right away.

Zoonotic Diseases: Giardiasis and salmonellosis are two examples of zoonotic diseases that tarsiers may harbor and spread to humans. It is imperative to maintain proper hand hygiene and clean the tarsier enclosure thoroughly in order to stop the disease from spreading.

Proper care and husbandry of pet tarsiers requires an understanding of their habitat and environmental requirements. Pet owners can guarantee the health, confinement are brought up by the choice to keep them as pets. Some people contend that the exotic pet trade endangers wild populations and undercuts efforts to protect tarsiers in their natural habitats, while others see pet ownership as a way to promote conservation awareness and the preservation of species.

Implications for Conservation: The poaching and trade of tarsiers for the pet trade may result in population decreases, habitat degradation, and heightened susceptibility to poaching and habitat loss for wild populations. In addition to advocating for the preservation of tarsiers in their native habitats, conservation organizations and animal welfare groups also support ecotourism and other initiatives that restore habitat.

Welfare Concerns: It may be difficult to meet the unique physiologic and behavioral requirements of tarsiers in captivity. Inadequate nutrition, confined living environments, and a lack of socialization can cause stress, behavioral problems, and decreased welfare in pet tarsiers. Providing proper care, enrichment, and veterinary attention to ensure the health and well-being of captive tarsiers is a prerequisite for ethical pet ownership.

Educational Opportunities: Those who support owning tarsiers as pets contend that doing so can offer the general public insightful teaching opportunities that will help them develop a greater understanding of biodiversity and the significance of conservation efforts. Caring pet owners can act as advocates for their species, bringing attention to the difficulties they face in the wild and the significance of protecting their natural environments.

2. Regulations and Legal Frameworks

At the international, national, and local levels, tarsiers are governed by an intricate web of laws and regulations regarding their ownership and possession. These laws differ greatly based on the species, region, and particular jurisdiction, so before obtaining a tarsier, prospective pet owners must become familiar with the applicable statutes and regulations.

International Conventions: A number of international conventions and agreements that regulate the trade in wildlife and work to conserve endangered species include provisions protecting tarsiers. Tarsiers are subject to international trade regulations and import, export, and re-export permits under the Convention on International Trade in Endangered Species of Wild Fauna and Flora (CITES).

National Law: A lot of nations have passed laws and rules that prohibit the ownership and trade of tarsiers inside their borders. These laws might forbid the taking and trading of wild tarsiers, impose ownership limitations, and require a license for the ownership of exotic pets. To guarantee compliance with legal requirements, prospective pet owners should investigate the laws and regulations that apply to tarsier ownership in their nation or region.

Local Ordinances: Municipalities and other local authorities may impose local ordinances and regulations in addition to federal laws governing tarsier ownership. Zoning restrictions, permit requirements, and prohibitions on bringing exotic animals into residential areas are a few examples of these regulations. In order to find out the legal requirements for owning tarsiers as pets in their community, potential pet owners should speak with the local authorities.

3: Advocacy and Responsible Ownership

Advocating for the welfare and conservation of these fascinating animals, adhering to legal requirements, and providing appropriate care are all part of responsible tarsier ownership. Owners of pets can actively encourage responsible ownership and support initiatives aimed at preserving tarsiers in the wild.

Research and Education: Before purchasing a tarsier, prospective pet owners should do extensive research on tarsier behavior, care, and husbandry requirements. To ensure the health and welfare of captive tarsiers, as well as to understand the responsibilities that come with pet ownership, education is crucial.

Regulation Compliance: Pet owners are responsible for making sure that their animals are owned in accordance with all local laws, rules, and permit requirements. This entails securing any licenses or permits required, abiding by zoning laws, and giving pet tarsiers suitable housing, attention, and enrichment.

Support for Conservation Efforts: By making donations to conservation organizations, taking part in ecotourism projects, and spreading awareness of the dangers facing wild tarsier populations, pet owners can help conserve tarsiers in their natural habitats. Through promoting conservation and ethical pet ownership, individuals can significantly contribute to the survival of these amazing animals for the enjoyment of future generations.

It takes careful thought and adherence to legal requirements and ethical principles to navigate the ethical and legal aspects of owning tarsiers. Pet owners can aid in the preservation of these fascinating primates and encourage responsible stewardship of our natural environment by being aware of the challenges involved in owning tarsiers and speaking out for their welfare and conservation. People can protect the welfare of pet tarsiers and contribute to the long-term sustainability of wild tarsier populations by educating themselves, abiding by the law, and supporting conservation efforts.

Chapter 4

Pet Tarsiers' Nutrition and Diet

In order to guarantee that pet tarsiers receive the vital nutrients they require to flourish in captivity, proper nutrition is crucial for their health and wellbeing. Due to their nocturnal lifestyle, tarsiers have specific dietary needs that must be satisfied with a diet that is well-balanced. We examine the nutritional requirements of pet tarsiers in this extensive guide, along with their natural feeding habits, suggested food sources, and methods for supplying a balanced diet in captivity.

1: Comprehending Tarsiers' Natural Diet

Tarsiers are primarily insectivorous in the wild, getting their sustenance from a diet of insects, spiders, and small vertebrates. Depending on their habitat and geographic location, their diet may vary, but common prey items include moths, small lizards, grasshoppers, and crickets. Tarsiers are expert hunters who locate and take down prey in the dark of the night by using their acute senses of hearing and sight.

Being opportunistic feeders, tarsiers eat a broad range of prey species to get the nutrients they need for development, procreation, and energy metabolism. Their diet high in insects gives them the vital protein, vitamins, and minerals they need to stay healthy and vibrant in the wild. To replicate their diet in captivity and guarantee the best nutrition for pet tarsiers, it is crucial to comprehend the natural feeding habits and dietary preferences of tarsiers.

2: Feeding Pet Tarsiers a Balanced Diet

It takes careful planning and consideration of the tarsiers' nutritional requirements to replicate their natural diet in captivity. Pet owners can guarantee a balanced and nutritionally complete diet for their tarsiers by providing a variety of food sources, even though it may be difficult to provide a diet exactly like that of their wild counterparts.

Live Insects: As they provide vital protein and nutrients, live insects ought to be the mainstay of a tarsier's diet while it is kept in captivity. Insects like roaches, fruit flies, mealworms, crickets, and waxworms are frequently available. To guarantee the best possible nutrition, insects should be fed a diet high in fiber and supplemented with calcium and vitamins before being given to tarsiers.

Commercial Primate Diets: For tarsiers kept as pets, commercially prepared diets designed especially for insects can be a practical and well-rounded choice. For tarsiers kept in captivity, these diets are usually supplemented with vitamins, minerals, and sources of protein to meet their specific nutritional needs. The tarsier's daily feeding schedule should include a premium primate diet that pet owners have chosen.

Fruits and Vegetables: Although in the wild tarsiers mostly eat insects, they occasionally add fruits, flowers, and plant materials to their diet. Small portions of fresh fruits and vegetables can be given to pet tarsiers as dietary supplements or as infrequent treats by pet owners. Fruits such as bananas, grapes, apples, carrots, and leafy greens are acceptable choices, provided in moderation to avoid stomach problems.

3: Feeding Timetable and Quantity

Feeding schedules must be established in order to provide pet tarsiers with the nutrition they require and to encourage good eating habits. Despite being nocturnal creatures, tarsiers have been known to feed all through the night and into the early morning. The following recommendations should be taken into account by pet owners when creating a feeding schedule for their tarsiers:

Frequency: To replicate their natural feeding behavior and guarantee a sufficient nutrient intake, pet tarsiers should be fed small, frequent meals throughout the night. By providing numerous feeding opportunities, multi-tarsier households can avoid overeating or food competition while allowing tarsiers to engage in their natural foraging behaviors.

Timing: The best times to feed are in the evening or at night, when the tarsiers are most active. In order to make sure that food is available when tarsiers are most active and likely to eat, pet owners can monitor their behavior and modify the feeding schedule accordingly.

Variety: Providing a range of food sources and textures promotes instinctive eating habits and offers chances for intellectual and enrichment stimulation. To offer dietary variety and avoid boredom, pet owners can alternate between various insect species, commercial primate diets, and fresh fruits and vegetables.

4: Supplementation and Nutritional Considerations

To guarantee nutritional sufficiency and avoid deficits, pet owners may need to supplement their tarsier's diet in addition to providing a variety of food sources. Important nutrients that might need to be supplemented include:

Calcium: For healthy bones and muscular function, tarsiers need an adequate supply of calcium. Calcium supplements can be given to tarsiers as a stand-alone supplement or as a powder or cuttlebone that is sprinkled over their food.

Vitamins: Extra vitamin supplementation may be beneficial for tarsiers, especially vitamin D3, which is necessary for the metabolism and absorption of calcium. Vitamin supplements are frequently added to commercial primate diets, but pet owners should speak with a veterinarian to find out whether their tarsiers need any extra vitamin support.

Protein: In tarsiers, protein is necessary for development, tissue repair, and general health. It is important for pet owners to make sure that their tarsiers are getting enough high-quality protein from commercial primate diets and insects.

Water: Keeping pet tarsiers properly hydrated is essential to their health and welfare. It is important to always have access to fresh, clean water, which can be simulated by natural dew collection through a drip system or a shallow dish. Tarsiers can get moisture from their diet, but for their general well-being and hydration, they must have access to clean drinking water.

The health and wellbeing of pet tarsiers depend on feeding them a diet that is complete and balanced in terms of nutrients. The specific nutritional needs of tarsiers kept in captivity can be met by pet owners by being aware of their natural feeding habits, providing a range of food sources, and making sure supplements are given as needed. Pet owners can support the health and vitality of their tarsier companions and create a lifetime bond based on mutual care and trust by carefully planning and taking into account their nutritional needs.

Chapter 5

Tarsier Socialization and Behavior

Tarsiers have a rich tapestry of behavioral adaptations shaped by their ecological niche and evolutionary history, which accounts for their captivating appearance and mysterious behaviors. In order to provide their tarsier companions a loving and stimulating environment, pet owners must have a thorough understanding of tarsier behavior and socialization. We examine the subtleties of tarsier behavior in this extensive guide, covering social dynamics, innate instincts, and methods for fostering mental and emotional health in captivity.

1: Examining the Natural Behavior of Tarsiers

Tarsiers are nocturnal primates that live alone and are well-known for their agility and keen senses. The demands of their forest habitat, as well as the difficulties of foraging, navigating, and avoiding predators in the dark of night, shape their natural behavior. Important facets of tarsiers' natural behavior consist of:

Tarsiers are nocturnal animals that hunt at night. They use their keen sense of hearing and large eyes to see through the darkness and find food. Their nocturnal habit is an adaptation to help them avoid competing with daytime predators and diurnal species.

Arboreal Adaptations: With long tails for jumping and balance between branches, grasping hands, and elongated fingers for gripping, tarsiers are exceptionally well-suited for living in trees. Because of their arboreal lifestyle, they can take advantage of vertical space in the forest canopy and obtain food sources that are inaccessible to species that live on the ground.

Living Alone: Although their home ranges may overlap with those of conspecifics, tarsiers are essentially solitary creatures that keep separate areas for mating and foraging. They use vocalizations, scent marking, and sometimes social interactions to communicate with other conspecifics, but they spend most of their time by themselves, hunting small vertebrates and insects.

2: Comprehending More Complex Social Dynamics

Tarsiers are solitary animals, but they display sophisticated social behaviors and communication techniques that are essential to their survival and procreation. When interacting with conspecifics, tarsiers exhibit a range of social interactions and affiliative behaviors, despite not being as gregarious as certain other primate species. The following are important tarsier social dynamics:

Behavior Related to Territory: Tarsiers are a highly territorial species that protect their exclusive home ranges from other members of their own species. Boundaries between territories are marked by scent marking, vocalizations, and sporadic violent confrontations with nearby people.

Social Interactions: When tarsiers come across conspecifics in the wild, they may get into social interactions with them. In contrast to more social primate species, these interactions are usually brief and rare. They may involve grooming, play, and mutual grooming.

Reproductive Techniques: To guarantee a successful mating and the survival of their offspring, tarsiers utilize a range of reproductive techniques. Usually, females give birth to lone children, who are raised by their mother until they are able to support themselves. Male tarsiers may engage in vocalizations, scent marking, and physical displays as a means of securing access to potential mates.

3. Encouraging the Mental and Emotional Welfare of Imprisoned Tarsiers

Encouraging the mental and emotional health of captive tarsiers requires providing a loving and stimulating environment. There are numerous tactics that pet owners can use to support their tarsier companions' natural behaviors, give them socialization opportunities, and foster cognitive development. Important factors to think about when fostering mental and emotional health are as follows:

Environmental Enrichment: In order to keep captive tarsiers from becoming bored and to encourage their natural behaviors, it is crucial to provide them with an engaging and diverse environment. Offering unique items and toys, building climbing frames and perches, and creating areas for foraging and exploration are a few examples of enrichment activities.

Opportunities for Socialization: Although tarsiers spend most of their time alone in the wild, they could gain from sporadic social interactions with compatible companions or conspecifics in captivity. Through supervised introductions, pet owners can create socialization opportunities for their tarsiers, giving them the freedom to interact and exhibit their natural behaviors in a safe environment.

Cognitive Stimulation: As highly intelligent creatures, tarsiers benefit greatly from puzzles and mental stimulation. By giving tarsiers puzzle feeders, hiding treats for them to find, and offering unique toys and enrichment items that stimulate manipulation and exploration, pet owners can help their pets' cognitive development.

4: Comprehending Tarsier Speech and Sign Language

The ability to vocalize is essential to tarsier communication because it lets individuals tell conspecifics about their presence, territory, and reproductive status. Even though tarsiers don't have as complex of vocalizations as other vocal primate species, they are nevertheless crucial for preserving social cohesiveness and communicating intentions within the group. Important vocalizations and cues for communication consist of:

Advertisement Calls: To let conspecifics know they are there and to mark their territory, tarsiers make loud, high-pitched calls known as advertisement calls. These calls, which are frequently made at night, may be used to ward off intruders or draw possible partners.

Contact Calls: When interacting with adjacent conspecifics, tarsiers may make quiet, chirping noises known as contact calls. During social interactions, these calls could be used to signal proximity or to stay in touch with other group members.

Agonistic Calls: During hostile interactions, tarsiers will make antagonistic calls to indicate submission, dominance, or territorial disputes. Depending on the individuals involved and the context of the interaction, the intensity and frequency of these calls may vary.

5: Behavioral Issues and Their Resolved in Tarsiers Under Captivity

Despite being hardy and adaptive creatures, tarsiers sometimes have behavioral issues in captivity that call for cautious handling and assistance from pet owners. Common behavioral problems in tarsiers kept in captivity could be:

Stress and Aggression: In captive tarsiers, alterations in their surroundings, social dynamics, or treatment may lead to stress and aggression. Owners should keep an eye out for telltale signs of stress in their tarsiers, such as vocalizations, aggression, or self-directed behaviors, and then use behavioral management techniques or environmental adjustments to address the underlying causes.

Reproductive Problems: In captivity, female tarsiers may face reproductive problems, such as trouble conceiving or raising young. It is recommended that pet owners furnish suitable nesting materials, privacy, and veterinary care in order to promote reproductive health and guarantee successful breeding results.

Creating a caring and stimulating environment for pet tarsiers requires a thorough understanding of their behavior and socialization. Pet owners can support mental and emotional well-being, encourage natural behaviors, and create enduring relationships with their tarsier companions by learning to recognize their natural instincts, social dynamics, and communication signals. By means of deliberate observation, enrichment of the surroundings, and socialization opportunities, pet owners can establish a happy and contented living space that promotes the well-being of their tarsier companions for an extended period.

Chapter 6

Veterinary and Health Care Considerations

Proactive veterinary care and management are necessary to guarantee the health and welfare of pet tarsiers. Even though these fascinating animals have a tough exterior, they can suffer from a number of health problems that can lower their quality of life in captivity. We examine preventive care, common health issues, and methods for fostering optimal health and wellness as we explore the vital medical needs and veterinary considerations for pet tarsiers in this extensive guide.

1: Pet Tarsiers: Preventive Care

Maintaining the health and well-being of pet tarsiers and delaying the onset of illness or disease require preventive care. Proactive pet care should include routine vaccinations, veterinary examinations on a regular basis, and parasite prevention techniques. Important elements of pet tarsier preventive care consist of:

Annual Wellness Exams: Qualified veterinarians with experience caring for exotic animals should conduct annual wellness exams on pet tarsiers. During these examinations, the veterinarian can evaluate the general health of the tarsier, keep an eye out for any indications of disease, and answer any queries or worries that the pet owner may have.

Fecal Testing: Regular preventive care for pet tarsiers should include a fecal test for parasites. Tarsiers may be prone to internal parasites like protozoa, tapeworms, and roundworms, which, if untreated, can negatively affect their health and general wellbeing. Fecal testing lowers the risk of complications and promotes gastrointestinal health in pet tarsiers by enabling veterinarians to detect and treat parasitic infections early.

Regular Vaccinations: Pet owners should speak with their veterinarian to find out if any vaccinations are advised for their tarsier, taking into account the animal's unique health status and risk factors, even though there may not be any vaccines specifically designed for tarsiers. Vaccinations against common infectious diseases, like rabies, might be advised based on the tarsier's interaction with other animals and possible infection sources.

Parasite Prevention: To shield their tarsiers from external parasites like fleas, ticks, and mites, pet owners should put in place a parasite prevention program. Maintaining the health and well-being of pet tarsiers may involve regular grooming, environmental sanitation, and the application of topical or oral parasite preventatives.

2. Typical Health Issues with Tarsiers as Pets

Pet tarsiers are prone to a range of illnesses and ailments that call for immediate medical attention as well as veterinary care. In the event that their tarsiers show any indications of illness or distress, pet owners should be ready to seek veterinary care. They should also become familiar with the warning signs and symptoms of common health issues in tarsiers. Typical health issues with pet tarsiers could be:

Respiratory Infections: Pet tarsiers frequently suffer from respiratory infections, especially when kept in conditions with inadequate ventilation or high humidity. Respiratory infections can cause breathing difficulties, nasal discharge, coughing, and sneezing. In order to avoid complications and facilitate recovery, it is imperative that pet tarsiers with respiratory infections receive prompt veterinary care.

Gastrointestinal Disorders: Stress, improper diet, or underlying medical conditions can all lead to gastrointestinal disorders in pet tarsiers, including diarrhea, vomiting, and gastrointestinal stasis. When there are any changes in a tarsier's appetite, fecal consistency, or behavior that could indicate gastrointestinal distress, pet owners should keep an eye on them and seek veterinary attention.

Dental Disease: Tarsiers frequently suffer from dental disease, especially if they are fed a diet high in sugar or carbohydrates. Dental issues in pet tarsiers, such as tooth decay, gum disease, and abscesses, can lead to pain, discomfort, and trouble eating. Regular dental care can help prevent dental disease and improve oral health in pet tarsiers. This care includes brushing their teeth on a regular basis and getting dental exams from a veterinarian.

Metabolic Bone Disease: Weakened bones and skeletal abnormalities are the outcome of metabolic bone disease (MBD), a disorder marked by a lack of calcium, vitamin D, or phosphorus. Pet tarsiers that are given an unbalanced diet or kept in settings with poor lighting or uneven temperature gradients may develop MBD. To stop MBD in pet tarsiers, pet owners should give them a well-balanced diet, the right supplements, and access to full-spectrum lighting or natural sunlight.

3: Emergency Preparedness and Veterinary Care

To guarantee that their tarsier companions have timely access to veterinary care and medical attention, pet owners should build a relationship with a qualified veterinarian with experience in caring for exotic animals. Apart from providing regular healthcare and managing common health problems, owners of tarsiers should be equipped to handle emergency scenarios and administer first aid in the event of an unexpected illness or injury. For pet tarsiers, essential elements of emergency preparedness and veterinary care include:

Emergency First Aid: In order to provide tarsiers with emergency care, pet owners should become familiar with the fundamental first aid procedures and techniques. This could entail tending to wounds, bracing fractures, and giving supportive care up until veterinary help arrives. It is recommended that pet owners keep a first aid kit stocked with necessary supplies and easily accessible contact details for emergency veterinary services.

Transportation and Housing: In case of an emergency, pet owners should have a safe and cozy enclosure or transport carrier on hand for taking their tarsiers to the vet. The carrier needs to be spacious enough to fit the tarsier comfortably during transportation, escape-proof, and well-ventilated. While their tarsiers are recuperating from disease or injury, pet owners should also make sure they have access to a safe and secure housing environment that satisfies their physiological and behavioral needs.

Veterinary Consultation: Pet owners should get in touch with their veterinarian right away to arrange for a consultation and examination for their tarsiers in the event of illness or injury. The veterinarian will evaluate the tarsier's health during the consultation, conduct diagnostic tests as necessary, and suggest the best course of action based on the tarsier's unique medical requirements and state of health. In order to aid in recovery and guarantee the wellbeing of their tarsier friends, pet owners should closely adhere to the advice given by their veterinarian and administer supportive care as instructed.

4: Enhancement and Behavioral Health

Ensuring the general health and well-being of pet tarsiers requires promoting behavioral health and well-being. Many enrichment plans and behavioral management approaches can be used by pet owners to encourage natural behaviors in captive tarsiers, keep them from becoming bored, and lessen their stress levels. Important factors to think about when fostering enrichment and behavioral health in pet tarsiers are as follows:

Environmental Enrichment: In order to encourage natural behaviors and keep pet tarsiers from becoming bored, it is crucial to provide them with a stimulating and varied environment. Offering perches, hiding places, climbing structures, and new toys or objects for exploration and manipulation are a few examples of enrichment activities. To keep their tarsier friends interested and engaged, pet owners should routinely switch up the enrichment items in their homes.

Opportunities for Socialization: Although tarsiers are mainly solitary creatures, they could gain from the occasional chance to socialize with like-minded captives or conspecifics. By introducing tarsiers under supervision, pet owners can create socialization opportunities that let them interact with each other and engage in natural behaviors in a safe environment. In order to ensure compatibility and prevent stress or aggression in their tarsier companions, pet owners should keep a close eye on their social interactions.

Cognitive Stimulation: As highly intelligent creatures, tarsiers benefit greatly from puzzles and mental stimulation. By giving tarsiers puzzle feeders, hiding treats for them to find, and offering unique toys and enrichment items that stimulate manipulation and exploration, pet owners can help their pets' cognitive development. To keep their tarsier companions interested and avoid habituation, pet owners should routinely switch up the enrichment activities they offer them.

In conclusion, it is critical to prioritize veterinary care and health care in order to protect the health and welfare of pet tarsiers. Tarsiers can have optimal health and longevity if their pet owners adopt a proactive preventive care approach, monitor for common health concerns, and provide timely veterinary intervention when necessary. With vigilant observation, emergency readiness, and proactive behavioral health and enrichment management, pet owners can create lifelong, satisfying, and harmonious relationships with their tarsier companions.

Chapter 7

Activities for Training and Enrichment

In addition to providing cognitive stimulation, social interaction, and a sense of fulfillment in captivity, training and enrichment activities are essential to the physical and mental health of pet tarsiers. Pet owners can create a happy and stimulating environment for their tarsier companions as well as strengthen the bond between them by participating in training sessions and offering enriching experiences. We examine the fundamentals of tarsier training and enrichment in this extensive guide, providing advice and tactics for creating a fulfilling and exciting environment for tarsier owners and their furry friends.

1: Recognizing Tarsier Training Principles

Positive reinforcement techniques are used in pet tarsier training to promote desired behaviors and discourage undesirable ones. Positive reinforcement is the process of praising, rewarding, or reprimanding undesirable behavior while ignoring or rerouting desired behavior. To keep the tarsier interested and motivated, training sessions should be brief and held in a peaceful, encouraging atmosphere.

Important training concepts for pet tarsiers consist of:

Patience and Consistency: Repetition, consistency, and patience are all necessary for tarsier training. When training their pets, pet owners should be patient and tenacious, rewarding desired behaviors on a regular basis and refraining from using punishment or unfavorable reinforcement.

Effective training requires clear communication, which is a prerequisite. In order to communicate with their tarsiers, pet owners should use consistent cues and signals. They should also promptly reinforce desired behaviors with rewards.

Good reinforcement: The cornerstone of tarsier training success is positive reinforcement. To encourage desired behaviors in their tarsier companions, pet owners can use rewards like favorite treats, praise, or playtime.

When training tarsiers, timing is everything. In order to effectively reinforce the association between the behavior and the reward, rewards should be given out as soon as the desired behavior is demonstrated.

Gradual Progression: As the tarsier gains proficiency, training should advance gradually, starting with simple behaviors and progressively raising the degree of difficulty. Complex behavior in pets should be broken down into smaller, more manageable steps, with rewards for each step that is completed.

2. Activities to Enhance Mental and Physical Stimulation

In order to prevent boredom, stimulate both the physical and mental faculties, and encourage natural behaviors in pet tarsiers, enrichment activities are crucial. There are numerous ways to enhance oneself, such as through social, cognitive, environmental, or sensory means. To keep their tarsier friends entertained and to enhance their general well-being, pet owners should offer a range of enrichment activities.

Enhancement of the Environment:

Enclosure Design: In order to encourage natural behaviors and offer chances for stimulation and exploration, it is crucial to design the tarsier enclosure to resemble their natural habitat. To promote movement and exercise, enclosures should have branches, perches, hiding places, and climbing structures.

Novel Objects: Adding toys and other unusual items to the tarsier enclosure can pique the animals' curiosity and present chances for exploration and manipulation. To keep their tarsier friends entertained and out of boredom, pet owners can provide them with objects like branches, ropes, puzzle feeders, and foraging toys.

Scent enrichment: To encourage exploration and investigation and to stimulate the olfactory senses, new scents are introduced into the tarsier enclosure. Pet owners can use commercially available scent products intended for enrichment or provide natural scent sources like herbs, spices, and non-toxic plants.

Enhancing Social Relations:

Social Interaction: Although tarsiers are largely solitary creatures, in captivity they might gain from infrequent social interactions with conspecifics or compatible companions. By introducing tarsiers under supervision, pet owners can create socialization opportunities that let them interact with each other and engage in natural behaviors in a safe environment.

Mirror Play: By enabling pet tarsiers to interact with their own reflection and partake in social behaviors like play, grooming, and territorial displays, mirror play can offer social stimulation and enrichment for these animals. To prevent stress or aggression in their pets, pet owners should supervise interactions and use a sturdy, safe mirror intended for enrichment.

Enhancement of Cognitive Function:

Puzzle feeders are interactive toys that put tarsiers through mental challenges and encourage foraging and problem-solving skills. Treats or food items can be concealed inside puzzle feeders by pet owners, who can then encourage their tarsier friends to manipulate the toy in order to retrieve the rewards.

Training Sessions: By offering mental stimulation and encouraging the acquisition of new skills, training sessions can act as cognitive enrichment activities for pet tarsiers. Positive reinforcement methods can be used by pet owners to teach their tarsier companions basic behaviors like recall, target training, and stationing.

Enhancement of Senses:

Audio Enrichment: You can stimulate your pet tarsier's auditory senses and give them auditory enrichment by playing natural sounds like bird calls, rainforest noises, or insect sounds. Owners of tarsier pets can provide a calming and stimulating auditory environment for their animals by using CDs, streaming services, or audio recordings.

Providing opportunities for visual stimulation and exploration within the tarsier enclosure is known as visual enrichment. To visually engage their tarsier companions, pet owners can hang colorful toys, provide visual barriers like curtains or foliage, or create visual interest with moving objects or patterns.

3: How to Train Your Pet Tarsier

For both pet owners and their tarsier partners, training a pet tarsier can be a gratifying and satisfying experience.

Pet owners can teach their tarsier companions a variety of behaviors and tricks that encourage mental stimulation, strengthen the bond between pet owner and tarsier, and improve the tarsier's overall quality of life by employing positive reinforcement techniques and clear communication.

Important tarsier training methods include:

Target training is teaching a tarsier to use their paws or nose to touch a designated target, like a stick or hand. To indicate the desired behavior, pet owners can use a clicker or verbal cue. Once the tarsier touches the target, they can reward them with a treat or praise.

Recall Training: In recall training, a tarsier is trained to respond to a name call or a particular cue. By rewarding the desired behavior with treats or praise, pet owners can use positive reinforcement techniques to encourage their tarsier to come to them when called.

Teaching a tarsier to stay in a specific area or on a particular perch or platform is known as stationing. To encourage a tarsier to remain in a desired location, pet owners can use positive reinforcement techniques. They can also use treats or praise to reinforce the behavior.

Agility Training: This includes teaching the tarsier how to maneuver through an obstacle course and carry out particular agility drills. To motivate their pet tarsier to finish the course, pet owners can install a range of obstacles, including tunnels, ramps, and balance beams, and employ positive reinforcement techniques.

4: Safety Observations and Safety Measures

Pet owners should prioritize safety and take precautions to ensure the well-being of their tarsier companions, even though training and enrichment activities can offer tarsiers many benefits. For training and enrichment activities, safety factors and precautions could include:

Pet owners should keep a close eye on training and enrichment activities to make sure their tarsier companions are safe and to avoid mishaps or injuries. In addition to offering opportunities for positive reinforcement and guidance during training sessions, close supervision enables pet owners to act quickly in the event that any concerns arise.

Training Environment: It is important to hold training sessions in a secure setting free from any potential dangers, such as electrical wires, poisonous plants, or sharp objects. During training sessions, pet owners should make sure the tarsier has a clear path to follow and remove any potential hazards from the training area.

Positive Reinforcement: To promote desired behaviors and discourage undesired ones, trainers should employ positive reinforcement strategies on a regular basis. Owners should refrain from using negative reinforcement or punishment on their pets because these methods can exacerbate anxiety or stress in tarsiers and interfere with their ability to learn.

Adaptation and Gradual Progression: To avoid overwhelming or frustrating tarsiers, training and enrichment activities should be tailored to each animal's unique needs and preferences and advanced gradually. To guarantee success and enjoyment, pet owners should keep a close eye on their tarsier's behavior during training sessions and modify the pace or difficulty level as necessary.

In conclusion, training and enrichment activities are critical to supporting pet tarsiers' physical and mental health as well as their sense of fulfillment in captivity, social interaction, and cognitive stimulation. Pet owners can create a fulfilling and enriching environment for their tarsier companions by using positive reinforcement techniques, offering a variety of enrichment experiences, and placing a high priority on safety considerations. This will strengthen the bond between the pet owner and the tarsier and promote overall health and happiness for both parties. Pet owners can create a lifetime relationship based on mutual enrichment, communication, and trust with their tarsier companions by being patient, consistent, and creative in their approach.

Chapter 8

Conscientious Ownership of Tarsiers

More than just giving food and shelter, responsible tarsier ownership entails recognizing the special needs of these fascinating animals and making a commitment to their mental, physical, and emotional well-being. Tarsiers need specific care and attention in order to flourish in captivity as exotic pets. We provide important guidelines for responsible tarsier ownership in this extensive guide, which covers topics like setting up a habitat, feeding, taking care of health issues, socializing, and legal considerations. By adhering to these rules, pet owners can support conservation and welfare efforts while guaranteeing their tarsier companions a happy and enriching life.

1: Recognizing the Needs for Tarsier Care

It's important to know the care needs of tarsiers and whether you can provide them before getting one as a pet. To ensure their health and well-being in captivity, tarsiers have particular needs that must be met in terms of diet, habitat, socialization, and veterinary care.

Nutrition and Diet: As insectivorous creatures, tarsiers eat mostly live insects like mealworms, grasshoppers, and crickets. They might also need to be supplemented with store-bought primate diets and the occasional fruit or vegetable. For their health, it is imperative to provide a well-balanced diet high in protein, vitamins, and minerals.

Habitat Setup: In order to replicate their natural habitat, tarsiers need a large, rich enclosure. This includes places to hide for privacy, places to rest and explore on branches and perches, vertical space for climbing, and environmentally enriching features like puzzle feeders and foraging opportunities.

Socialization and Enrichment: Although tarsiers are solitary creatures, they might gain from infrequent social encounters with like-minded captive companions or conspecifics. Furthermore, for their mental and emotional health, enrichment activities like toys, puzzle feeders, and sensory stimulation are essential.

Health Care: Tarsiers need routine veterinary examinations, prevention of parasites, and fast medical intervention in the event that they exhibit symptoms of disease or trauma. To address their specific medical needs, it is imperative to locate a veterinarian with experience in caring for exotic animals.

Legal Considerations: It is important to learn about the laws governing the ownership of tarsiers in your community before obtaining one as a pet. Some legal jurisdictions may limit or outright forbid the keeping of tarsiers as pets, while other legal jurisdictions might call for licenses or permits.

2: Providing Tarsiers with an Appropriate Habitat

For the sake of the health, comfort, and general wellbeing of pet tarsiers, an appropriate habitat must be created. As much as possible, the enclosure should mimic their natural habitat, offering chances for climbing, exploring, and exhibiting their natural behaviors. Important factors to set up a habitat are as follows:

Size of Enclosure: Tarsiers need a large enclosure that allows them to jump, climb, and explore. The tarsier should be able to move around freely in the enclosure, which should be big enough to hold several perches, branches, and hiding places.

Vertical Space: Being arboreal creatures, tarsiers spend a significant portion of their lives in trees. In order to promote natural behaviors, the enclosure should have vertical space for perching and climbing, with branches and platforms positioned at various heights.

Environmental Enrichment: In order to keep pet tarsiers from becoming bored and to encourage mental stimulation, environmental enrichment is crucial. Creating opportunities for foraging and exploration, offering novel objects and toys, hiding treats, and installing puzzle feeders are a few examples of enrichment activities.

Substrate: The tarsier enclosure's substrate needs to be secure, pliable, and clean. Paper-based bedding, coconut fiber, or specialized substrates for reptiles are examples of suitable substrates. Steer clear of substrates that could irritate the respiratory tract or be consumed.

Temperature and Humidity: Tarsiers need a stable environment to survive because they are sensitive to changes in both of these parameters. The enclosure should be kept between 75 and 85 degrees Fahrenheit (24 and 29 degrees Celsius) and between 50 and 80% relative humidity. It might be required to use humidification and heating equipment to create and preserve these conditions.

Lighting: In order to maintain vitamin D synthesis and control their circadian cycles, tarsiers need access to either natural or artificial lighting. If tarsiers are kept indoors, full-spectrum UVB lighting might be required to give them enough UVB exposure, especially if they don't get natural sunlight.

3: Satisfying Tarsier's Nutritional Needs

The health and wellbeing of pet tarsiers depend on feeding them a diet that is complete and balanced in terms of nutrients. Tarsiers primarily eat live insects, but they can also benefit from occasional fruits and vegetables as well as commercial primate diets. When making food choices for tarsiers, it's important to take into account:

Live Insects: As a primary source of protein, vitamins, and minerals, live insects ought to be the mainstay of a tarsier's diet. Insects such as mealworms, roaches, crickets, and waxworms are frequently offered. Before being given to tarsiers, insects should be fed a diet high in nutrients and supplemented with calcium and vitamins.

Commercial Primate Diets: For tarsiers kept as pets, commercially prepared diets designed for insects can be a practical and well-rounded choice in terms of nutrition. To satisfy the special nutritional needs of tarsiers kept in captivity, these diets are supplemented with vitamins, minerals, and sources of protein.

Fruits and Vegetables: Although in the wild, tarsiers mostly eat insects, they occasionally add fruits and vegetables to their diet. Small portions of fresh fruits and vegetables can be given to pets as treats or dietary additions on occasion. Bananas, grapes, apples, carrots, and leafy greens are good choices.

Supplementation: To guarantee adequate nutrition, tarsiers might need extra supplements, especially calcium and vitamin D3. A veterinarian should be consulted by pet owners to establish the proper supplementation schedule for their particular tarsier, taking into account its unique nutritional requirements and overall health.

Water: Keeping pet tarsiers properly hydrated is essential to their health and welfare. To mimic natural dew collection, fresh, clean water should always be available in a shallow dish or via a drip system. Tarsiers can get moisture from their diet, but for their general well-being and hydration, they must have access to clean drinking water.

4: Guaranteeing Veterinary Care and Health for Tarsiers

Pet tarsiers need to be taken care of with preventive care, frequent veterinarian checkups, and fast medical attention if they exhibit any symptoms of illness or injury. Important factors for the health and veterinary care of tarsiers are as follows:

Annual Wellness Exams: Qualified veterinarians with experience caring for exotic animals should perform annual wellness exams on Tarsiers. These examinations offer the chance to evaluate the tarsier's general health, keep an eye out for any disease or illness-related symptoms, and respond to any worries or inquiries from the pet owner.

Fecal Testing: Regular preventive care for pet tarsiers should include a fecal test for parasites. If left untreated, internal parasites like roundworms, tapeworms, and protozoa can have a negative effect on a tarsier's health. Fecal testing lowers the risk of complications by enabling veterinarians to detect and treat parasitic infections early.

Regular Vaccinations: Pet owners should speak with their veterinarian to find out if any vaccinations are advised for their tarsier, taking into account the animal's unique health status and risk factors, even though there may not be any vaccines specifically designed for tarsiers. Vaccinations against common infectious diseases, like rabies, might be advised based on the tarsier's interaction with other animals and possible infection sources.

Parasite Prevention: To shield their tarsiers from external parasites like fleas, ticks, and mites, pet owners should put in place a parasite prevention program. To reduce the likelihood of parasitic infestations, routine grooming, environmental hygiene, and the application of topical or oral parasite preventatives may be advised.

Emergency Preparedness: In the event that their tarsiers become ill or are injured suddenly, pet owners should be ready to handle emergencies and administer first aid. This could entail keeping a first aid kit fully supplied, being able to provide basic first aid, and having access to emergency veterinary care.

Fifth: Encouraging Lawful and Moral Tarsier Ownership

Ownership of tarsiers that is responsible takes into account welfare, conservation, and legal requirements. The health of their tarsier companions should come first for pet owners, but they should also respect the environment and any local laws pertaining to tarsier ownership.

Conservation Awareness: The illegal pet trade, habitat loss, and deforestation pose serious threats to tarsiers, so advocacy and conservation efforts are essential to their survival. By spreading knowledge, giving to conservation groups, and refraining from obtaining tarsiers as pets from the wild or from unlicensed sources, pet owners can aid in the conservation of tarsiers.

Legal Compliance: Pet owners should learn about and comprehend the local laws pertaining to the ownership of tarsiers before obtaining one as a pet. Some legal jurisdictions may limit or outright forbid the keeping of tarsiers as pets, while other legal jurisdictions might call for licenses or permits. To ensure that they are legally and morally the owners of their tarsier companions, pet owners should abide by all applicable laws and regulations.

Outreach and Education: Responsible tarsier ownership, conservation, and welfare can be greatly enhanced by pet owners teaching others about these topics. Pet owners can encourage others to become activists for tarsier conservation and welfare as well as ethical and responsible pet ownership practices by sharing their knowledge and experiences with tarsiers.

Proper care of tarsiers necessitates a dedication to addressing their special requirements and advancing their conservation, health, and well-being. Through the provision of an appropriate living environment, adherence to dietary needs, veterinary care, and ethical and legal guidelines, pet owners can foster a fulfilling and enriching bond with their tarsier companions while simultaneously aiding in their conservation and welfare. Pet owners can make a significant contribution to guaranteeing tarsiers have a bright future through advocacy, education, and compassionate care, both in the wild and in captivity.

Chapter 9

In conclusion, would you be a good home for a tarsier?

Choosing to bring a tarsier into your home as a pet is a big decision that needs careful thought, taking into account the tarsier's special needs, your lifestyle, and your capacity to give it the care and attention it needs. In our final section, we go over important factors to help you decide if a tarsier is the right pet for you.

Recognizing Tarsier Requirements:

Tarsiers are fascinating animals that differ from more popular pet species in that they have special needs and traits. Because they are native to Southeast Asia, tarsiers are arboreal primates with unique dietary, environmental, and social requirements that must be satisfied to maintain their health and welfare in captivity.

Dietary Requirements: As insectivorous creatures, tarsiers mainly consume live insects like mealworms, grasshoppers, and crickets. To guarantee a balanced and nutritionally complete diet, they might also need to be supplemented with commercial primate diets and occasionally fruits and vegetables.

Environmental Requirements: Tarsiers need an expansive, well-furnished enclosure that reflects their natural surroundings and offers chances for climbing, exploring, and mental stimulation. To encourage natural behaviors and avoid boredom, the enclosure should have branches, perches, hiding places, and environmental enrichment features like puzzle feeders and sensory stimulation.

Socialization: Although tarsiers are largely solitary creatures, in captivity they might gain from sporadic social interactions with conspecifics or compatible companions. Providing them with enrichment and socialization opportunities can improve their mental and emotional health.

Veterinary Care: Tarsiers need routine examinations by veterinarians, as well as preventive care and quick medical attention in the event that they exhibit symptoms of disease or injury. To address their specific medical needs, it is imperative to locate a veterinarian with experience in caring for exotic animals.

Legal Considerations: It is important to learn about the laws governing the ownership of tarsiers in your community before obtaining one as a pet. Some legal jurisdictions may limit or outright forbid the keeping of tarsiers as pets, while other legal jurisdictions might call for licenses or permits.

Evaluating Your Way of Life:

Think about how a tarsier's needs fit into your schedule, way of life, and living arrangement before bringing one into your house. To suit their specific needs and create a suitable living environment, tarsiers demand a major time, energy, and resource commitment.

Time Commitment: Tarsiers need daily care, which includes enrichment activities, cleaning, feeding, and socializing. They may also survive for 20 years or longer in captivity, necessitating a sustained commitment to their care and welfare.

Space Requirements: In order to engage in their natural behaviors of climbing, jumping, and exploring, tarsiers require a large, well-enriched enclosure. Make sure your house has enough room for the tarsier's enclosure and enough room for it to receive both mental and physical stimulation.

Financial Responsibilities: Taking care of a tarsier requires regular maintenance, veterinary care, setup costs for the enclosure, and food. Owning a tarsier can be expensive. To guarantee your tarsier companion a happy and fulfilling life, be ready to make financial investments in their care and well-being.

Compatibility with Other Pets: If you already have other pets, think about how a tarsier will fit into your home's dynamics. It's possible that tarsiers are incompatible with dogs, cats, or other small animals, and that having them around could stress out or aggravate current pet relationships.

Evaluating Your Skills:

A tarsier's complex needs must be met with knowledge, skills, and resources that come with caring for them. Evaluate your capacity and willingness to provide your tarsier companion the care and attention they require to ensure their health and well-being.

Knowledge and Experience: Before getting a tarsier as a pet, familiarize yourself with the care needs, behavior, and husbandry techniques of this species. To develop your knowledge and abilities in caring for tarsiers, think about volunteering at a wildlife sanctuary or obtaining experience working with exotic animals.

Patience and Commitment: Tarsiers are special and complicated creatures that need care and well-being to be provided for with patience, consistency, and commitment. Be ready to devote the necessary time and energy to fostering a trustworthy bond with your tarsier companion and resolving any issues that may come up.

Access to Resources: Make sure you have access to resources like a licensed veterinarian with experience caring for exotic animals, trustworthy vendors offering dietary supplements and live insects, and trustworthy sources of information regarding the upkeep and care of tarsiers.

Making the big decision to bring a tarsier into your house as a pet involves giving careful thought to a number of variables, such as the tarsier's particular requirements, your lifestyle, and your capacity to give it the care and attention it needs. Tarsiers are not suitable for everyone, but for committed and knowledgeable pet owners, they can make fascinating and rewarding companions.

Before getting a tarsier as a pet, spend some time learning about their needs, evaluating your lifestyle and skills, and determining whether you can provide for them. A tarsier might be the ideal pet for you if you can guarantee veterinary care, provide a suitable habitat, adhere to dietary needs, and follow the law.

However, it might be best to look into other pet options or consider supporting tarsier conservation efforts through advocacy, education, or volunteer work if you are unable to commit to the responsibilities of tarsier ownership or if their needs do not fit with your lifestyle.

Responsible tarsier ownership ultimately entails putting these fascinating animals' health, welfare, and well-being first while cultivating a rewarding and enriching bond built on mutual respect, trust, and understanding. You can start a fulfilling lifelong journey of companionship and discovery with your tarsier companion by making well-informed decisions and devoting yourself to their care and well-being.

Chapter 10

Commonly Asked Questions (FAQs) regarding Ownership of Tarsiers

- What exactly are tarsiers, and where are they found?

Small, nocturnal primates with large eyes and distinctive physical characteristics are called tarsiers. They are indigenous to parts of Southeast Asia, which includes Malaysia, Indonesia, and the Philippines.

- Are tarsiers good pets for all people?

No, not everyone should keep tarsiers as pets. To maintain their health and well-being in captivity, they have particular care needs that must be satisfied in terms of food, habitat, socialization, and veterinary attention.

- What legal ramifications come with keeping a tarsier as a pet?

Depending on the jurisdiction, different laws have different requirements and regulations regarding the ownership of tarsiers. Owning tarsiers as pets may be restricted or outlawed in some areas, while obtaining a permit or license may be necessary in others. Before getting a tarsier as a pet, it is imperative that you learn about and comprehend the local laws.

- What food is best to give my pet tarsier?

The main food source for pet tarsiers is live insects like mealworms, grasshoppers, and crickets. To guarantee a well-rounded and nutritionally complete diet, they might also need to be supplemented with commercial primate diets and occasionally fruits and vegetables.

- How much room do tarsiers require in their habitat?

In order to replicate their natural habitat, tarsiers need a large, well-stocked enclosure with plenty of space for climbing, exploring, and mental stimulation. The tarsier should be able to move around freely in the enclosure, which should be big enough to hold several perches, branches, and hiding places.

- Is socialization with other animals necessary for tarsiers?

Although tarsiers are largely solitary creatures, in captivity they might gain from sporadic social interactions with conspecifics or compatible companions. Nonetheless, it's critical to keep a close eye on social interactions to make sure everyone gets along and avoid tension or hostility.

What medical attention are tarsiers in need of?

Preventive care, timely medical attention in the event of illness or injury, and routine veterinary examinations are necessary for Tarsiers. To address their specific medical needs, it is imperative to locate a veterinarian with experience in caring for exotic animals.

- In captivity, how long do tarsiers live?

When given the right care and handling, tarsiers can survive in captivity for up to 20 years or longer. Their longevity and well-being can be enhanced by providing a suitable habitat, satisfying their dietary needs, making sure they receive veterinary care, and attending to their social and behavioral needs.

- Like other pets, can tarsiers be trained?

Indeed, positive reinforcement methods can be used to train tarsiers to reward desired behavior and punish undesirable behavior. To keep the tarsier interested and motivated, training sessions should be brief and held in a peaceful, encouraging atmosphere.

- How can I contribute to tarsier conservation? Are they an endangered species?

Because of habitat degradation, deforestation, and the illicit pet trade, tarsiers are regarded as vulnerable or endangered species. Supporting conservation groups, educating people about tarsiers' predicament, and avoiding tarsiers as pets that have been illegally or wild-captured are all ways you can contribute to their conservation. You can also volunteer or make donations to organizations that defend the populations and habitats of tarsiers."

www.ingramcontent.com/pod-product-compliance
Lightning Source LLC
Chambersburg PA
CBHW050324230526
45471CB00005B/2334